the enchanted
orchid

the enchanted orchid

max fulcher

NEW HOLLAND

First published in Australia in 1999 by
New Holland Publishers (Australia) Pty Ltd
Sydney • Auckland • London • Cape Town

14 Aquatic Drive Frenchs Forest NSW 2086 Australia
218 Lake Road Northcote Auckland New Zealand
24 Nutford Place London W1H 6DQ United Kingdom
80 McKenzie Street Cape Town 8001 South Africa

National Library of Australia
Cataloguing-in-publication data:

Fulcher, Max, 1935- .
The enchanted orchid.

Includes index.
ISBN 1 86436 535 8.

1. Orchids - Pictorial works. 2. Orchids. I. Title.

635.9344

Publishing General Manager: Jane Hazell
Publisher: Averill Chase
Commissioning Editor: Derelie Evely
Designer: Andrew Cunningham – Studio Pazzo
Editors: Margaret Kelly and Julie Nekich
Project Coordinator: Julie Nekich
Reproduction: Colour Symphony, Singapore
Printer: South China Printing

Page 2: Jade Buddha and a green *Cymbidium*.
Following pages: *Dendrobium farmerii* in a giant clam shell.

contents

Once upon a time   8

Why men grow orchids   16

Colour me blue   26

Bottoms up!   36

How to keep your orchids happy   46

The ornamental orchid   56

Airborne orchids   66

The princess and the showgirl   76

I am a camera   86

Cattle-lay-ya, don't even think it   94

The lost slipper   102

The tiger and the dancing ladies   106

Moon orchids   112

The noble Hawaiians   122

Pathway to the flowers   132

Index   144

once upon a time

Not so long ago, a young boy stood
on the edge of a vast rainforest. A stiff wind blew as he looked
up into the tree tops and saw a dazzling cascade of buttery
yellow flowers waving to him from the branches.

'What are they Dad?'

'They're orchids. King orchids.'

Page 8: An extreme close-up of the King Orchid or Rock Lily, *Dendrobium speciosum*.
Above: A forest of *Dendrobium speciosum*, Stony Range, Dee Why, Sydney.

Orchids on wind-swept mountains were my passport to a whole world of flowers. There's a special magic about orchids. Like ravishing birds of paradise, they burst before your eyes to vanish and suddenly reappear in another part of the forest. They can be frivolous and funny, or sexy as hell. Perfumed, gaudy, seductive or rather straight-laced. Orchids fit the occasion. You may recall your first encounter with the gorgeous royal of the flower kingdom. Like oysters, crusty French bread, and salt-rimmed margaritas, orchids are an acquired taste.

I had a love–hate relationship with orchids during those dreadful teen years in the '50s when boys in tuxedos took girls in frilly ballgowns to formal dances. Sweating it out for days beforehand. Making that first phone call to arrange the date. Meeting her family for the first time. Fronting fathers you didn't know. Thrusting forward a be-ribboned orchid corsage in a see-through celluloid box. Feeling the urge to bow ever-so-slightly as you did so. Watching to see if the mother pinned it on the dress (they liked it) attached it to the girl's wrist (they hated it) or, worse, affixed it to your date's beaded bag (to disappear under a pile of rabbit stoles and hand-crocheted cashmeres on arrival at Cloudland, the dance destination).

Orchids were central to the whole dating ritual. The fact that the floral tribute cost a couple of guineas – half a week's pay packet – was beside the point. Let's not even begin to discuss those massive bridal bouquets (with innocent-looking white orchids wired into the arrangement) that were tossed to the next lucky bride-to-be as the happy couple headed off to their honeymoon and wedded bliss in his FJ Holden.

> You don't have to stop loving roses to enjoy orchids. You no longer have to be a millionaire to grow them.

Gathering stories for this book, I read over some of the orchid articles I've accumulated over the years. Certain words jumped out at me as I scanned the stories. *Tempting, alluring, tantalising, desirable, erotic, slinky, svelte, intriguing, exotic, provocative, racy, pungent, raunchy, sensual, pleasure seeking, sexual, sensuous, vulgar, suggestive, lusty, delicious.* The interesting thing is that most of the pieces were written by women, whereas the earlier diaries and accounts of orchid collectors were mainly written by men. The language of the eighteenth and nineteenth centuries was much more circumspect, descriptions more discreet — but the inference was the same. It's just that a more sedate set of phrases was used to describe the unveiling of the ravishing maiden.

Here's how Charles Darwin addressed the delicate subject of mating orchids in his 1877 epic, *The Various Contrivances by which Orchids are Fertilised by Insects*: 'Nature has endowed these plants (Catasetum) with, for the want of a better word, sensitiveness, and with the remarkable power of forcibly ejecting their pollina to a considerable distance. Hence, when certain points of the flower are touched by an insect, the pollina are shot forth like an arrow, not barbed however, but having a blunt and excessively adhesive point.'

Bullseye, Mr Darwin.

This is a book about the love of orchids — the rare and the beautiful now within our reach. You don't have to stop loving roses to enjoy orchids. You no longer have to be a millionaire to grow them. Travel orchid pathways. Search out moon orchids in hotel foyers. Visit orchid houses in botanic gardens. Watch for orchid shows and meet orchid fanciers. Buy a big pot of flowering orchids and watch them blossom indoors for weeks on end.

Armchair traveller? Then lie back and enjoy the ride. These stories are true. The photos are taken in natural light. No fakes, no filters. But be warned — if you're looking for a highfalutin' dossier full of orchid hocus-pocus, you'd better buy another book. This one is about experiencing joy, love and the beauty of nature — and spreading it around.

Opposite: Yamamoto Yellow Ribbon 'Delight', Hawaii.
Following pages: Renanthera Brookie Chandler.

why men grow orchids

'What is it about orchids — their exotic sensuality?

their infinite variety? — that lures men to collect them

with a passion reserved for no other flower.'

Some years ago a friend of mine forwarded an article from American *Town & Country*. The subject was orchids, and why men grew them. The lead copy read: 'What is it about orchids – their exotic sensuality? their infinite variety? – that lures men to collect them with a passion reserved for no other flower.'

Good question, because the thought has bugged millions. What is the relationship between a man and his orchis – almost to the exclusion of all else in the plant kingdom? The magazine article was simple, easy to read. The inference being that orchid growers toyed with the sex life of their darlings, prodding here, stroking there, creating rare and strange beauties under purple grow lights over Park Avenue. True. Orchids, men, and sex are eternally intermingled. Though today, with a good percentage of orchid societies presided over by women, poor blokes are beginning to run out of places to call their own.

Skulduggery and treachery flourished in the great orchid hunts before the turn of the century. Lies, deceit, imprisonment, disease and death.

By and large, I gleaned three points from the informative article:

(1) Carl Linné, a Swedish botanist better known in the scientific community by his Latinised name Caroulus Linnaeus, chose the name 'orchis' when classifying the plant kingdom way back in 1753. 'Orchis' is the Greek word for testes. And the shape of the roots of the terrestrial orchids evidently reminded Linnaeus of the male testicle. The Linnaean system is an artificial or sexual method of botanical classification that Linnaeus formulated.

(2) Skulduggery and treachery flourished in the great orchid hunts before the turn of the century. Lies, deceit, imprisonment, disease and death. Boys will be boys and more of that later.

(3) Growing orchids is not so esoteric and scientific. More, it's a matter of discipline. Getting up early. Spraying them. Coaxing them along. Sustained attention is what makes growth.

Unfortunately, in her haste, my friend neglected to rip out page 217 of the article and it took 17 years and persistent searching to dig out a back copy to see how the story finished. Apart from the disclosure that Japanese aristocracy treasured orchids as one of the group of four noble plants; and that they grew *Dendrobium moniliforme* as the flower to make men live a long life, there was not a lot to report. And somewhere around five hundred years ago in Yucatan, Emperor Montezuma drank a liquid ground from orchid fruits and cacao seeds before battle. Aztecs believed the drink made them tireless gladiators. It was a concoction of

Page 16: *Phaius tankervilliae*, Far North Queensland, Australia.
Opposite: Prize-winning red Vandas, Philippines.

chocolate, vanilla flavoured with ground pods from *Vanilla planifolia* – and that's what the Spanish conquistadores took back to Europe. An orchid and chocolate smoothie.

The last quote in the story holds true: 'Every time I see a new species – and they are still being found – I get as excited as I did with the first orchid I ever had. That's how orchids are: magical, and constantly stimulating because the subject is infinite. You can never have them all.'

Even with page 217 still firmly in my grasp, I'm not that certain what the male attraction to orchid growing is all about. Sure, I have mates who tickle and titillate various bits of flowers in the interests of cross fertilisation. And I'm sure they have a jolly time. But even as I look up the dictionary meaning of the word 'flower', I find explanations such as 'the sexual reproductive structure of the angiosperm usually consisting of gynoecium, androecium and perianth'. No, we're not going that far in. Let's just adopt a plant, enjoy watching it grow and bringing the flowers to blossoming radiance, with the full intention of helping the plant to the next season when hopefully it will perform again.

Getting back to **skulduggery and treachery**. The accounts of what men went through in the jungles to drag home the prize is well documented. I have a favourite I would like to share. It's from a March 1923 copy of *Sea, Land and Air* magazine, a noted Australian journal of the day: 'The dangers of the collector's task are terrible. Eight

naturalists seeking various specimens in Papua once dined at Port Moresby, and in one year after there was but a single survivor. Even this favoured person was terribly afflicted, for, after a sojourn in the most malarious swamps, he spent twelve months in the Townsville Hospital, and left without hope of restored health. Two collectors seeking a single plant died one after the other of fever. A collector detained at Port Moresby went far inland to look for an orchid he had heard of and the natives brought him back from the swamps to die. A French collector who insulted a native chief because his men would not guide him to a spot where certain orange and black orchids were said to be growing, was clubbed to death ...'

All this in a little epic called 'The Price Men Pay, Perils of Orchid Collecting' by Frank Reid.

Orchids grow from the tropics to the Arctic Circle – sea level at the Equator and as high up as 12 000 feet. In Vietnam for instance, orchids are part of the currency. Meet someone with a name like Mai Lan and the orchid is ever present, Lan meaning Orchid in Vietnamese. Once you start on the orchid trail you'll find the flower flourishing everywhere.

The Chinese hold orchids as a symbol of loyalty. *Bletilla*, a ground orchid native to China, is the preference of herbalists for its bulbs gathered in the wild. This has led to the amethyst-purple flower becoming a very rare find in the wild. It is unclear to me what medicinal value bletillas might possess, so be wary if you experiment with a flowering pot

Opposite: The hard cane *Dendrobium antennatum*, New Guinea.

Above: A forest nursery, Chang Mai, Thailand.

of the foxy little fellow picked up in your local nursery or flower market.

In medieval Europe, legends existed about wonderful potions and cure-alls made from orchids. Marco Polo may have been slipping *Bletilla* (or something other than silk and spices) up and down the Silk Road. Simon and Garfunkel made parsley, sage, rosemary and thyme more the go in the herbarium. And (sorry fellas) there is not one skerrick of truth in the rumour that little blue Viagra pills contain the essence of Orchidaceae. Stay off the orchids, get with fresh vegies and take long walks.

Above: The Cooktown Orchid *Dendrobium bigibbum* is the floral emblem of Queensland.
Following pages: *Blue Vanda* orchid house, Chang Mai, Thailand.

colour me blue

'Near the village of Larnac, oak woods are passed in which *Vanda coerulea* grows in profusion, waving its panicles of azure blue in the wind. We collected seven men's loads of this superb plant for the Royal Gardens at Kew, but owing to unavoidable accidents and difficulties, few specimens reached England alive.'

Page 26: Vanda hybrid, Singapore.
Above: This Vanda needed heated glasshouse conditions to bloom in Sydney.

The association of flowers with music is fairly universal. Cherished moments always seem to be punctuated with symphonic crescendo and a profusion of blossom – or have I been seeing too many old MGM movies lately? On the subject of blue orchids, it's the silence that surrounds me. The sighting of a blue orchid is a memorable event. The moment stands still.

The rarest and, for me, the only blue orchid is a species called *Vanda coerulea*. A true blue. It grows in India, Assam, and other romantic-sounding places we're inclined to associate with stories of the early Raj.

Vandas are monopodial, meaning single-footed. Unlike many other orchids, they climb upwards and are not rooted to the ground by a big clump of pseudo bulbs and such. They have strappy-looking leaves that alternate from either side of the main stem. This gives the plants an art deco appearance, and allows them to scoot up trees using long fleshy roots that spring from between the leaves, and grab for the next foothold as the plant soars towards sunlight. Vandas love sun. They come in pungent colours like ginger and sandalwood, cinnamon pinks and honey tans. Except that sublime blue number, *Vanda coerulea*.

The evasive blue orchid was first seen by Dr William Griffith in the Khasia Hills of Assam in 1837. Brought to England, it was flowered and lost. Mid century, Sir Joseph Hooker recounts in his Journal of his meeting in the same hills of Assam and the fate of the prized vanda: 'Near the village of Larnac, oak woods are passed in which *Vanda coerulea* grows in profusion, waving its panicles of azure blue in the wind. We collected seven men's loads of this superb plant for the Royal Gardens at Kew, but owing to unavoidable accidents and difficulties, few specimens reached England alive.'

Other obsessed collectors ran around raping forests and downing every huge vanda-enhanced tree in sight, denuding newly uncovered forests of their free-flowering epiphytes. The local government had to call a halt as the mania gripped wealthy enthusiasts, and wild blue vandas faced imminent extinction. Sir Joseph did get one thing right: '... now this (Assam) winter's cold, summer's heat and autumn's drought, and above all this constant free exposure to fresh air and the winds of heaven, are all things which we avoid exposing our orchids to in England. It is under these conditions that *coerulea* best grows.'

It's unfair to solely blame the English. Rich Americans were also into pillage. And the French. And the Germans. The misinformation (still held by many today) that all orchids need hot, steamy growing conditions led to the demise of all but a fraction of the plants that were hauled back to the hothouses of the rich and famous.

As a young executive in the '70s, I had all the pretensions of that era. Moon orchids on the baby grand, blue orchids in the hallway. I installed a small glasshouse in the backyard of my inner city residence (the vision was ludicrous) and started to grow 'hothouse' orchids. Lustrous milk-white phalaenopsis for the grand piano, radiant vandas for the entrance hall – the full catastrophe.

The orchids thrived. Moon orchids everywhere. Some of the warm-toned vandas tried to push their way through the glass-house roof (if you're going to grow vandas under glass, build a reasonably tall structure). Everything bloomed freely except (guess) the elusive blue *Vanda coerulea*. Not a beep.

Blue vandas grow in mountainous foothills, and while they need warm days, they also require cool nights . . .

Capricorns like to do things the hard way, so I decided to shoot off to Burma and discover what I was doing wrong. The answer? Blue vandas grow in mountainous foothills, and while they need warm days, they also require cool nights before deciding to break into full song.

Back to Sydney, out of the hothouse came the rare species (and it had been hard to buy) to be strapped on to a handy Bangalow palm.

The orchid seemed to favour outdoors in Sydney. The plant took a couple of years to reach a few feet in height, but the roots clung tightly to the trunk of the palm, and we

scattered on some bark and ashes to humour the little fellow.

In the late winter of the third year, a nubby spike appeared. I was speechless. Everyone arrived to watch the slow progress as the spike grew to form eight or nine buds. We kept peering into the heart of the buds, trying to ascertain just what shade of blue our *coerulea* would reveal.

The first bud broke – a wonderfully soft, shy blue. We set up. Camera at the ready. Next day, the continuing promise of another unfolding flower.

On the morning of the third day, we looked out from the kitchen window right into the blazing red bottom of a friendly neighbourhood bul-bul, head down chomping vanda buds. All gone.

No photo opportunity. Worse, the Bangalow threw down a couple of heavy fronds, knocked the vanda off the tree trunk and smashed the plucky little plant to smithereens. We tied it back up, but the end was in sight. When you decide to grow heavenly blue orchids, don't overheat the glasshouse, and ban bangalows and bul-buls.

These days, blue orchids are best viewed by visiting vanda growers where, with any luck, you'll meet one or two prize specimens in bloom. Rarely for sale. Always expensive. Or go visit the hills around Chang Mai above Bangkok in Thailand where blue orchids are a dime a dozen. There in orchid garden upon orchid garden you'll see hybrids that have been crossed with warmer coloured vandas. Blues in every imaginable colour. Azure,

Above: Blue Vanda hybrid, Mandai Gardens, Singapore.

cobalt, cerulean blue, indigo, sky, navy, prussian, lavender blue, deep purple. *Deep Purple*, now there's another story.

Some people talk to their plants, and I often sing to mine when I'm in a good mood. They seem to respond. Somewhere between *Stardust*, *Skylark* and *Deep Purple*, Hoagy Carmichael penned another popular love song:

> *I dreamed of two blue orchids . . .*
> *blue orchids*
> *only bloom in your eyes.*

The lyrics of *Blue Orchids* made a big hit in World War II circles because it was also the nickname for the Air Force. My father was in the Air Force. As the flavour of the month in their stepping-out blue rig, the air boys were highly rated at nightclubs and cocktails by the silver fox set – but some artist got it wrong at the publisher's. The two orchids depicted on the cover of the Chappell sheet music were cattelayas – good pinks and whites, but never convincing blues. The swing set around the world – New York, Sydney, London, Paris – played in blue spotlights as smoke got in their eyes, but there were no blue vandas in the picture. Yet who cared! Not Hoagy. Not Ava. Not Rita. And most certainly not Dietrich, film's legendary *Blue Angel*.

Opposite: *Vanda coerulea*, photographed in the foothills above Mandalay, Burma.

Following pages: *Stanhopea nigroviolacea* bursting with energy and blatantly vanilla-scented.

bottoms up!

Now here's an orchid that could make a maiden blush.
The zany subtropical *Stanhopea* dangles great clusters
of blossom from its nether regions, down below
the pseudo bulbs and large leafy foliage.

What first seems like a goofy quirk of nature turns out to be one of the magnificent miracles of nature. Growing stanhopeas is one of the most fascinating experiences in the plant world. Overwhelming, viewed from beneath, advanced specimens can throw two, three, four and more spikes within a period of a few days. It's common to see masses of forty to fifty flowers in one grand sweep.

It must be marvellous to happen upon these dizzying apparitions in the jungles of South America – where jungle still remains. Reports of pillage and slash-and-burn operations set off alarm bells. That many wild orchids now exist only through the good graces of botanic gardens and dedicated growers is hardly tribute to man's passage through the kingdom of plants. But it is also important to keep a positive picture of these alluring plants swaying in tropical jungle atmosphere when you set about growing them.

Stanhopeas must never be housed in solid-sided pots (how would the flowers break through the terracotta?) and are best grown in wire baskets and viewed hanging high in your garden or fernhouse. They don't need full sun as much as, say, vandas or Indian dendrobiums, so if big brown sunspots break out on the leaves, your plants have either a fungus or a sun cancer. If you live near the beach, don't dry out stanhopeas in salty drafty winds, for heaven's sake!

Joy of joys, stanhopeas bloom late in the season after most else has given up the ghost. Late spring into early autumn, the sturdy stanhopea keeps popping away. And I do mean 'pop'. So vigorous is the flowering, that on the morning of flowering, you can hear buds break out into bloom if you get up close and personal. Children love the soft 'pop'. They also love the pong.

With the pops come exotic and extremely erotic perfumes that call the birds and the bees – a plus for the romantically inclined and young at heart. But don't linger in wonderment too long, the fragrance is short-lived. A day or two into the event, the original sweet smells sour to a cross between bitumen and old boots. Not exactly unpleasant, but warn friends before they sidle up too close. As I said, stanhopeas are not the most politically correct creations of nature.

One of the most common *Stanhopea* species is also quite vulgar. Large. Flashy. Fleshy. Smelling first of vanilla and then dead meat is the *Stanhopea nigroviolacea*. It was quickly pointed out by an experienced orchid-growing mate that most of the plants you see in home gardens are wrongly named *Stanhopea tigrina*. Tiger stripes are a whole different kettle of fish. *Stanhopea nigroviolacea* (think negro and violets) is profusely purple and liver, splashed on a creamy-yellow base. *Tigrina* is a little more sophisticated to my eye and considerably more rare. It's best to get the opinion of an expert if you are in doubt.

In the slow-moving days of sail ships, botanists pressed specimens of exotic flowers for future reference. The bulky, almost luminous stanhopeas did not travel well pressed in the pages of a book. Finally,

Page 36: *Stanhopea wardii* flourishes out-of-doors in inner-city Sydney.
Opposite: Ice-green *Stanhopea inodora*.

Stanhopea tigrina flowered in England around 1839 and *Stanhopea nigroviolacea* a few years after that; the latter being recognised as a variety of *tigrina* before it was finally given species status, a point overlooked by many subsequent botanists. The issue is being debated by a number of dedicated collectors as they publish articles and share experiences through the marvels of e-mail and the internet.

Stanhopeas are fairly husky and need dividing quite regularly. To this end, you can cultivate a fair number of orchid plants for not a great outlay – providing you are prepared to invest the years of effort. Remember, plant in loose chunks of small bark or a mix of bark and cymbidium compost, and make sure the lining of the baskets is thinner than usual so the spikes can push through. Keep the rhizomes above the bark level or you'll choke the stanhopea to death.

Here is a short list of varieties that grow easily in temperate outdoor conditions. Happy hunting!

Stanhopea nigroviolacea: This fellow seems all-male to me. Big, robust, easy growing. The giant flowers hang down in almost embarrassing profusion. This is the orchid that first-timers look sideways at and don't know what to say. They blush.

Stanhopea inodora: My favourite and a more diminutive mate to *nigroviolacea*, if you believe opposites attract. No shy daisy, this ice green beauty is a hardy grower, but with a delicate countenance. Each time a photograph of *inodora* is published, I receive calls from hobbyists wanting to buy some bulbs or swap for a piece of their *nigroviolacea*. Not possible, because it took eight years before my sole specimen was ready to divide. Now there are three baskets, but I'm loath to let any of my fair ladies out of sight in case the mother plant expires. We were four. I did give one blooming plant away in a fit of heroics last summer. I think about my absent inodora even as I write this chapter. It's 3 am and the urge to ring the new owner is compelling because it's spring and time to start a little light fertilising.

Above: *Stanhopea graveolens.*
Opposite: Detail, *Stanhopea graveolens.*

Stanhopea reichenbachiana: Porcelain white and spicy sweet perfumed, this species is found in hot and almost inaccesible rainforests of Colombia. A rare find, here photographed with Barney Greer, author of *The Astonishing Stanhopeas*, my bible as a reference on the genus.

Stanhopea wardii: A sunny yellow deckled with purplish flecks, this fellow is a late finisher. Blooms late in summer. In Sydney, it hangs easily in trees or under shadecloth – anywhere there is a bit of shelter. Often confused with *S.graveolens*, but the distinguishing mark on *S.wardii* is the dominant dark eye at the centre of the flower.

Stanhopea graveolens: Cool growing (bush houses, glass houses, shaded patios, garden balconies protected from the wind), this golden beauty hails from Mexico and Honduras. You'll note the bright orangey lip without *S.wardii* dark eye spot. Baskets of the healthiest plants I have ever seen simply hung from the branches of an accommo-dating avocado tree in Paddington, Sydney. The slightly off-colour scent of fresh bitumen exuded by the day-old flowers quite enhanced the bouquet of a couple of glasses of Aussie white. 'Bottoms up!'

Top: *Stanhopea reichenbachiana*, Colombia.
Above: Bernard (Barney) Greer, author of *The Astonishing Stanhopeas*.

Opposite: *Stanhopea wardii*.
Following pages: *Coelogyne pandurata*, Malaysia and eastern Asia.

how to keep your orchids
happy

'Orchids are tenacious little buggers …
start with something easy. Just make sure the roots
don't wobble around. No grip, no grow.'

There's a pecking order in the orchid world. Top of the tree sit botanists who can discuss, in infinite detail, the biology of each genus. Then follows an important group of sub-botanists. Dedicated people who know the ropes, but sometimes get the official facts wrong because they just speak from their own experience. Next the doers and the growers. The show people. The dedicated fanciers. The fourth and fifth levels are where most orchid enthusiasists dwell. I'm about level four, aiming for two. As a mate of mine in Western Australia puts it, 'I buy orchid plants smothered in flowers and buds. There's something flashy and cheap about an orchid absolutely smothered in flowers. A bit over the top, but isn't that the very nature of the orchid? Over the top.'

Here are some maxims gathered en route. You may find them helpful, and they will stop you going barmy as you start getting serious about orchids. First advice – **learn by doing**. Read only the bits set in bold type if you're just passing through enjoying the view.

'Orchids are tenacious little buggers.' The best advice I ever received on orchid growing came from a very experienced grower as he demonstrated how to tie a native orchid to a piece of old tree trunk. It's easy. 'Just make sure the roots don't wobble around when you're finished. Wobbly roots mean the orchid can't get a secure grip. No grip, no grow.'

Start with something easy. Practise on bromeliads. They're hardy and thrive in similar conditions to orchids, being air, tree and ground dwellers. Pineapples are bromeliads and if you stick the spiky head of a pineapple into the ground (flesh bit down), it will grow and like-as-not bear fruit, even in

Page 46: Australian *Sarcochilus hartmannii* with a Balinese bronze deer.

Top: Vigorous young orchid roots in a slatted wood basket.

Above: Heads down, tails up. Orchids are a full-time job.

Above: Label it and date the label. You'll know what you're talking about at a later date.

temperate zones. Naturally, the warmer the climate the bigger the fruit.

Easy orchids to begin with are Australia's *Dendrobium kingianum* and *Dendrobium speciosum* (mentioned above), the creamy-yellowish orchids previously called King orchids in northern parts of Australia, and rock lilies in the south because they grow in clumps on pinkish Sydney sandstone. Same orchid genus, different growing habits. All you need are small chunks of bark and some lumps of sandstone. Don't bog down their roots in garden soil. Let the roots wander and breathe. And if you can't beg, borrow or steal a hunk of dendrobium from a shop or grower mate for under twenty dollars, then they've seen you coming. Twenty-five dollars tops! Start hunting or give in and just enjoy the pages of a good picture book of orchids. Don't feel guilty. Not everyone is cut out to garden and there are any number of ways to appreciate orchids.

Orchids love light and air just like you and me. Get them into sunny garden spots.

It's good to see what you're getting. Beginners should start with mature plants. I'm inclined to buy plants in bud and bloom. Reputable growers can usually be trusted, but mix-ups have been known to occur.

The flowering plant comes at a premium price but, hey, that flowering head is a real value-added plus. And you get to see where you should be heading in the next flowering cycle.

Sunny side up. We tend to hide precious objects and possessions in deep vaults and dark secret places. A mistake with the queen of flora. Orchids love light and air just like you and me. Get them into sunny garden spots. Don't fry, grill, bake or boil them in strong sun. Just give them a little more light and air if you want to see them blossom. Or settle for spectacular green leaf plants with fascinating root systems if you give them too much shade. Half the fun is watching plants grow – blooms are really a bonus.

Label it. If your plan is to own more than half a dozen plants, then get into the habit of ticketing each one from the very start. You know at a glance what you are growing. The species name, if that be the case: its lineage if crossed with another orchid. Always date the plant with year and month each time you re-pot it. Nurseries can supply little plastic strips for the exercise. Also, you'll need a waterproof pen. A German Staedtler works for me. Just don't leave it in the pocket when your shirt goes to the laundry.

Make sure they can pass water. Orchids hate wet feet. The bottom of the rainforest might be damp and full of lobbies and leeches, but it's leaf matter, twigs and porous rocks that low-growing orchids thrive on. Like other epiphytes (*epi*, above or on; *phyte*, plant) they have also found a way to live above the mass of stifling undergrowth.

They have learned to cling to trees in the aerial garden up top in the forest and it's not unusual to see orchids blooming right on the thatched roofs of village houses as you scoot around the tropics. But orchids are not parasites. They do not take nourishment from their hosts. Orchids collect humus and insects, and have a porous outer layer on their root system that absorbs rain and mist quickly. Between drinks (they aren't soaks so don't overwater) the stems and leaf systems house moisture to eke out the dry spells.

Up in the forest canopy, it's bits of bird poo, air-carried elements that nourish the orchid's world. Imitate that environment and let the water run off, just like it does in a rainstorm. Replace the bird droppings with regular dousings of **weak** organic fertilising solutions. I use the old organic mixes rather than the new salts, but I'm certain you'll get other advice from the next gardener you meet.

Rest in peace. Don't feed orchids during winter. They're sleeping so do not disturb until springtime. As soon as those little green tipped orchid roots shoot, recommence feeding. When you mix it, halve the density it says on any label. Half as much, twice as often. Orchids like to be regular. **Little and often**, all through summer.

Orchids hate cement. It's the alkaline lime content in the cement mix that screws things up. Don't try to grow dendrobiums directly onto house brick walls or around cement pathways. And don't put broken cement in the bottom of orchid pots for drainage and ballast. Recycle broken terracotta pots, lumps of sandstone and such. **Orchids love sandstone**, so keep a few bits about tennis ball size to grind over the plant wherever you think it. Little green-tipped roots thrive on it.

... epiphytes need a home with an atmosphere that replicates their natural habitat ...

Orchid roots seek the light, so punch holes in pots. Slash, thrust, burn, bore or cut. Do anything you can think of to help access light, air, and water to the root systems of *Dendrobiums*, *Vandas*, *Phalaenopsis* – in fact, all the epiphytes. You might not like the holey pot look, but epiphytes need a home with an atmosphere that replicates their natural habitat, and air is what those roots search for. Hey, whose bed is this anyway?

Spend your money on good healthy plants and never complain about price. My first orchid purchase was *Dendrobium gracillimum*, a natural hybrid cross between *Dendrobium speciosum* and its cousin *Dendrobium gracilicaule*. It was growing in a piece of old industrial piping (ah, for the days when plumbers used terracotta drainage pipes), and no one else bid for it at the auction. It went for a song and has bloomed every year for 15 years for a two-week period. In full bloom it travels indoors and makes a great talking piece – the first sign that winter has had its chips.

A basket case. This slatted arrangement is popular in the Orient. Except Asians make it in a pulpy wood that rots all too quickly. Longer lasting are durable little hardwood knockoffs (see photo above) available in orchid centres if you're quick. They disappear from stockists' shelves at the blink of an eyelid. Better still – find a handyman or do it yourself. The wooden slats bolt together (square or hexagonal as you wish) and the container is then wired to hang like any other basket. Great for all the *Aerides* tribe including *vanda* and *phalaenopsis* and air-growing *dendrobium*.

A recycled white plastic bag makes a temporary orchid house tied overhead a single plant when weather turns inclement. Clear plastic is not the way to go – it allows sun to burn through like a magnifying glass and scorch the leaves. A couple of stakes keep the sides of the bag away from the subject and, while the white hoods look a bit Ku Klux Klan, the idea works well enough. The dog will try to cock his leg, of course.

Above: Hanging slatted baskets give orchid roots breathing space.

Staking. Because of their heavy pendent flowering habit, many orchids (particularly cymbidiums) need a life support or they will snap as the blooms wax to full beauty. Where stakes are needed, set them as the flower spikes are forming and therefore more malleable. Just be careful you don't break the tender bracts because orchids do not just spring another stem. Once lost, a year's wait.

Pantyhose tie up. It's the give and take of old pantyhose that make them ideal running mates for tying epiphyte orchids to trees and old fencing posts. Cut circlets of fabric from the legs and, linking them daisy-chain fashion, you can make lengths as long as you like. The hose expands as the tree grows, whereas tight wires cut into the trunks. Invisible nylon fishing line is also a handy stand-by for the orchid-growing gardener.

A ready reckoner. A handy pocket book on orchids should go along for the ride when you start your orchid travels. My golden guide to orchids is called just that, *A Golden Guide to Orchids*, and the few dollars it cost me have been returned in kind many times over. I'm on my third copy because wear and tear on the book has been enormous. Suitcase, camera case, coat pocket. The subjects in the book are species and naturally occuring plants are the best place to start. Maybe the orchid on your dining room table is a big colourful hybrid. But it is helpful to know the parentage of offspring. It's all in the genes, remember.

Reading botanical names. Carl Linné (or Caroulus Linnaeus, as referred to in the chapter Why Men Grow Orchids) simplified

the classification of living things enormously.
What may seem at first complicated to the
novice becomes workable when following
Linnaeus' system of naming, ranking and
classifying plants and animals. He gave
one Latin name to designate the genus
and another to declare the species.

Hence the name *Dendrobium chrysotoxum*
reveals that in the tree growing *Dendrobium*
group (dendro = tree) there is a species
orchid that flowers golden yellow, called
chrysotoxum (chryso = gold, toxum = arch).
When hybrids are created they are usually
named by the hybridiser. This part of the
identification carries an initial capital letter to
avoid confusion with the species name which
is always written using a lower case initial.

Top: A happy mix of orchids and tropicana.

Above: *Dendrobium kingianum* thrives in an easy-to-fashion hollow log.

Following pages: Orchids surround dancers in a traditional garden at Sanur Beach, Bali.

the ornamental orchid

Ikebana was introduced into Japan from China.
It was originally created to decorate statues of the Buddha.
Modern, minimalist, it often uses orchids as
the pivotal point of the arrangement.

When boy meets girl in Bali, flowers follow. Gifts; tokens; offerings; celebrations—the ornamental orchid everywhere.

We decided to make an orchid picture in the traditional Balinese garden at La Taverna on Sanur beachfront. If you think Baz Luhrmann's *Romeo and Juliet* was full of colour abstractions, look again to our Romeo and Juliet on pages 54 and 55.

First the boy meets the girl. Then the housekeeper arranges the decorative orchids. The hair and make-up artist. The gardener to loop the coconut leaves. Sandy, in charge of transport, will 'fire the candles'. And one cameraman. Candles! Camera! Action! The game is in play and Dorothy Lamour and Bing Crosby could not have had as much fun on *The Road to Bali*.

Orchis belongs to the new age. A flower for all seasons …

A princess royal. A winged butterfly skimming far horizons. Wind-blown grasses. Lost loves. Balmy tropical nights. California dreaming. Crashing waves. Orchids in the moonlight. Small wonder men telegraph their innermost feelings through orchid tributes. The woman is beautiful, the flower is beautiful. The message soon follows.

Ikebana, the Japanese art of flower arrangement, is an art form that whets appetites with feelings for nature. Eastern floral arrangement is a philosophy greatly admired in Western society. Modernist, minimalist, it often uses orchids as the pivotal point in the arrangement. The language of Ikebana is both primitive yet highly sophisticated. It mirrors nature. The objective is simplicity, a oneness, but the actual technique is complex. As with iris, citrus and cherry blossom, the Japanese demonstrate a definite affinity with orchids, often expressed in haiku, a form of Japanese verse developed in the sixteenth century. Traditional haiku contains seventeen syllables and celebrates joyful nature as do many English sonnets. A famous writer of haiku was a monk called Basho. His first pseudonym was Tosei, and he lived in the Basho-Tosei heritage at the end of the seventeenth century. His ability to bring sudden awareness to impact on simple acts of nature is electrifying – in this instance he describes a favoured flower.

Orchid – breathing
incense into
butterfly's wings.

As wilderness disappears, new orchids are still being discovered. Fascination with this intriguing member of the herbaceous family grows and unconquered, the orchid may manifest as the ultimate flower emblem of the new millennium. As the romance of medieval roses reaches its zenith, and with much already written about daisies, daffodils and

Page 56: Orchids and wild grasses on a Chinese prayer table.
Above: Ripe pawpaw, tangelo and orange-red *Laelia cinnabarina*.

hillsides of heather, the ubiquitous orchid (like Thursday's child) has far to go.

Red poppies, white carnations, French fleur-de-lis and willowy bamboo patterns are icons of bygone eras. Orchis belongs to the new age. The Age of Aquarius. A flower for all seasons, with a past, present and future. Express yourself – with orchids.

Orchids bring a sense of occasion to just about every Asian dish.

Table manners. Orchids make very agreeable table companions. A spray of Singapore orchids separates to supply a single flower per guest per setting. Or the whole spray looks devastating draped beside a communal bowl of steaming rice. Every visitor to Asia has seen such simple table arrangements. But, back home, we forget to follow suit. Here are a few friendly reminders.

Orchids for breakfast. A tropical fruit breakfast at Noosa, Queensland. Orange *Laelia* decorates great wedges of deep golden pawpaw with tangerine tangelos. Who needs a single rose on a tray? **Orchids for lunch.** A single white flower unites seared tuna with cucumber, green papaya, chilli and lime salad. **Orchids for supper.** White nobile orchids perched on a plate of moneybags. Orchids bring a sense of occasion to just about every Asian dish. Just don't toss them in your spring salads.

Flower salads may be the go, but leave it to nasturtium and calendula. The heavenly orchid tastes like hades.

Stamp collections. Orchids feature time and again on the colourful stamps of just about every Asian and Pacific nation. A series was struck for the sixth Asia Pacific Orchid Conference held in Townsville, Queensland, September 1998. A joint issue between the Commonwealth of Australia and the Republic of Singapore. Australia selected endemic Cooktown orchids and the Moth (moon) *Phalaenopsis*. Singapore chose the Bamboo and Tiger orchids.

Born travellers. Orchids decorating the bathroom positively transform the atmosphere (but don't leave them there for days on end). Better still, flower them in one garden position, then move them in full bloom to a more eye-catching position. They are great little garden fillers. A bush house full of ferns and greenery presents a new facet the minute a swinging basket of orchids is added. Outdoor table settings take on a life of their own with a pot of orchids as a centrepiece. Friends of mine take their orchids along in their campervan on holidays and weekends away – especially if the plants are in bud. Wouldn't want to miss a moment of the action!

Figure of fashion. The orchid represents a perennial part of Paris fashion shows. Chanel. Dior. All the great houses have used expressions of the orchid at some point during their career. Young British designer Alexander McQueen staged a dazzling

Top right: Delicious Indonesian moneybags and *Dendrobium noblie* (Alba form).
Bottom right: A seared tuna embellishment with *Cattleya Little Angel*.

scenario for his fourth *haute couture* show for Givenchy at the 1998/99 Paris Collection. Through a rainforest backdrop, replete with rushing waterfall, stepped a scintillating model in an orchid strewn full-length evening dress, cloaked in a jacket of grasses, reeds, feathers and rushes.

The engaging flowers broadcast messages at once sexy and sensual – overflowing with intrigue.

An old clam shell makes a fantastic home for flowering dendrobiums. Time-worn craggy clams juxtapose the pinks and yellows of pendulous dendrobiums. The plants can be flowered in one place and displayed in another. Fool friends. Bring orchids indoors when the flowers are at peak. They'll think you grow them there all the time. A single spray triggers untold emotion. Although not all orchids are perfumed, the merest hint of their subtle scent ignites fantastic feelings – choose carefully. The brush of orchids across a cheek is positively titillating. Try it. The engaging flowers broadcast messages at once sexy and sensual – overflowing with intrigue.

Above: Orchid-strewn Paris fashion.
Opposite: Joint stamp issue to coincide with the 6th Asia Pacific Orchid Conference, 1998.
Following pages: Detail, *Dendrobium pierardii*.

airborne orchids

Dendrobiums regenerate through the air.

They're born travellers and go wherever the wind blows.

Silver Knight streaked home in the 1971 Melbourne Cup and I was on cloud nine. What a horse! If I remember, it won at 8 to 1.

Rather than a ritzy dinner for two (in those distant days any small windfall bought the moon and stars in an up-market eatery), I blew the money on a gorgeous golden orchid called *Dendrobium densiflorum*, and gave it to the girl who had so successfully translated my pre-cup dream about invincible knights in shining armour. I wanted a silvery orchid, but golden yellow was as close as the grower could get at short notice. Three stunning cascades of brilliant yellow. What a vision!

We decided to try to bloom it again the next year on her balcony. No luck. Not enough sunlight hours. Next year 'Silver Knight' travelled to her family home – a sunny spot on Sydney's North Shore. We hung it under a large spreading frangipani. Still no go. Too much shade.

Finally, I agreed to take the poor thing back to the fold in inner-city Darlinghurst. Such a jewel, struggling for life in derelict Darlo. Well it worked! More sun. More light. A lot less care and attention. The Knight divided time and again to make further generations of Knights that hung high in the slatted bush house. They did not need a glass house, continuing to bloom every October until the neighbour's trees closed in and stole their light.

Dendrobium densiflorum is one of the Indian dendrobiums, and flowers just a little later than his Australian family members. Dendrobiums are epiphytes (meaning air growers) and that's how they regenerate. Through the air. They're born travellers and go wherever the wind blows.

Love of dendrobiums (*dendro*, tree; *bios*, life) became my driving passion. Orchids led me to photography, for instance. Indian dendrobiums called me to Asia. Observing how they flourished in their natural habitat – Burma, then Thailand and Malaysia – the South China Seas. I didn't go rushing off into the jungle to find rare species. You didn't have to. The plants were best observed in little remote villages where locals treated them with simple respect.

They seemed to grow on nothing but air, and what brilliance! Undernourished to Western eyes, but flourish they did on bits of wood and old charcoal.

In Burma, a two-hour ride by jeep took me to a small village up back of Mandalay. A few modest dwellings grouped around community cooking pots and such. It was after sunset, and a meal was being prepared. Smoke billowed from the fires, kids played and gathered around. Someone showed me their special orchid when suddenly a little three-year old started to scream and run. What a

Page 66: Detail, *Dendrobium farmeri*.

Above: *Dendrobium chrysotoxum* translates as the Golden Arch orchid.

A burst at dawn. A volley at dusk. You're never quite sure when kingianums will bombard the garden with their wonderful fragrance.

racket. What an embarrassment. The people smiled politely and Rose (the guide who drove me in an old jeep to this off-limits area) just laughed and explained, 'He is three. He has never seen a white man before. You are their first visitor. You are bigger than he expected.'

Above: *Dendrobium pierardii* silhoueted in first light before the Shwedagon pagoda, Rangoon.

The markets in Rangoon and Mandalay were an easier photographic target. Clusters of yellow orchids hanging overhead and poking out of jam tins with holes punched in the sides – or bits of old pots. They seemed to grow on nothing but air, and what brilliance! Undernourished to Western eyes, but flourish they did on bits of wood and old charcoal. Here's a list of easy growers:

Soft cane, open air growers

Dendrobium kingianum: An Australian native. Pink, magenta, white and all the in-betweens. 'Kingys' (there's that abbreviated slang again) are much loved little national treasures. Marvellously well-perfumed in the main, they'll suddenly scent a night-time garden to outdo the jasmines. A burst at dawn. A volley at dusk. You're never quite sure when kingianums will bombard the garden with their wonderful fragrance.

Dendrobium speciosum: Buttery yellow and standing on spikes as thick as a child's wrist, this beauty can produce a bunch of fifty to one hundred small flowers on one stem. It grows on sandstone rocks or graces tree tops in Queensland where great clumps nestle between the fork and the trunk, often displaying a dozen and more spikes in the one burst of Spring. The variety *hillii* is a softer cream, near white.

Dendrobium nobile: Something of a Black-eyed Susan effect. Pinky-mauve flowers, with a deep magenta throat on a creamy-white background, *nobile* stands tall but can make a wonderful hanging basket effect. Bracts

Above: *Dendrobium Ellen.* A popular, highly perfumed hybrid.

with up to a dozen flowers crown the plant and make a great showpiece. More modest, though no less eye-catching, is the white (*alba*) variation. Then the high-voltage hybrids in every imaginable hue and variation. Bright yellow, soft pink, gold and buttery orange – plus a heavenly host of soft gelato colours. I've included a whole chapter on *Nobile*-type orchids cloned in Hawaii.

Hard cane, plenty of warmth and sunshine

Dendrobium chrysotoxum: 'Indian' dendrobiums like a dry spell to flower. If your winter climate is also rainy, bring this fellow (and the rest of his family) indoors for the cool spell – but keep the light. Do not feed the plant while it is hibernating. Water once a week? Of course. But – repeat – NO fertiliser. Expect bright yellow flower spikes with an even deeper yellow throat.

Dendrobium farmeri: Clusters of translucent pink and yellow blossom cascade from the plant in late spring. It grows well in the sunny part of a bush house, or hang it in the branches of some accommodating, fairly open tree. Native to Burma and Nepal.

Dendrobium densiflorum: An extraordinary display of sunny yellow flowers hang in lantern-like clusters, up to eighty a stem. Each flower is about 30–40 cm across and the lip is covered in a velvety down. Native to Nepal and Assam, the plant grows to about 40 cm and thrives in morning sunlight.

Dendrobium pierardii:This is one of the orchids I went to visit in Burma. I shot it in silhouette right in front of the golden Shwedagon pagoda. My plant in the bush house back home had trails of little pink flowers over a metre long when I bought it in bloom. Each year it bloomed with a shorter stem until one year, phffutt. This is not an orchid for the absolute beginner. But what a prize when it works!

Hot house baby

Dendrobium bigibbum: Popularly known as the Cooktown Orchid, this fellow needs heat, and if you look on the map you'll see why. Yep. That's the tropics. The deep reds of Queensland's floral symbol are used in colouring many hybrids. It's long-lasting and sits well in posies and corsages. Somewhat insulting use for this fiery little native of northern Australia and New Guinea. Needs hothouse conditions in temperate climates to be seen at best advantage.

Above: Detail, *Dendrobium bigibbum*, the Cooktown Orchid.
Opposite: Hard cane New Guinea Dendrobium, Queensland.
Following pages: Jade cymbidiums.

the princess and the showgirl

Grace Kelly wrapped in Monroe's sex appeal.

Cymbidiums are real show-stoppers. Think

Spice Girls meet La Stupenda.

If you want to see Snow White, Sleeping Beauty, Mary Poppins and Minnie Mouse, go to Disneyland. But if you've an urge to meet the Red Baron, Fire Lips, Peggy Sue and Goldfinger, get to an orchid show and search out cymbidiums.

The names orchid breeders give their offspring are worth the price of admission alone. Jake's Folly, Dolly X Claude (X = crossed with), Showgirl 'Glamour Jane' and Sleeping Nymph. Beats horse racing!

Asians and Americans adore these cool growers, but the genus is almost taken for granted in Australia. In season, you see cymbidiums on sale in just about every florist shop, plant nursery, flower stall, church fair and supermarket chain. Swank shopping malls flood their fashion areas with potted orchids as do many up-market eateries. Look into half the gardens in Australia and somewhere in half shade under a gum tree or standing prettily in amongst spring annuals, you'll find a clump of *Cymbidium*, doing nicely, thank you.

The story is that in World War II, England (Kew Gardens by most accounts) shipped her genetic stock to Australia for safe-keeping – not so much to avoid the bombs as to escape fuel rationing. Energy was diverted to the war effort, as cymbidiums were apparently shipped to Australia where climate made growing conditions easier.

Not wishing to pass up a good story, it seemed a bright idea to verify the tale for inclusion in this book. But (leaving hillsides of turned stones in the attempt) I retreat.

Exactly **who** started this rumour? From one orchid show to another; grower upon grower; one exhibitor to the next; it was the same story. Enthusiasts gave the idea currency while sending you to the next person in the cymbidium chain. 'Yes. That would be right. I've heard that. Why don't you speak to … ' In Asia, you learn not to point when you ask directions. Locals just nod in the affirmative and send you off in whatever direction you happen to be pointing. Could it be the same with cymbidium fanciers?

Phone calls, email and internet, foot slogging, and still no published evidence of the happening was located. Approaches to large botanic gardens were neatly side-stepped and now, as patriotic as the whole thing sounds, I am forced to admit that it might be dealing in myths and legends to pass the story along to you. But it is such a wonderfully romantic notion. 'Australians save their mother country's cymbidiums as part of the *Win-the-War* effort.' I can almost see the headlines and live in constant hope that one morning, upon opening the desk top, e-mail (maxful@nsw.bigpond.net.au) will affirm a thrilling piece of history that maybe never was. Meantime, it's the Camelot of cymbidiums.

Once you get the climate correct, orchids are not such difficult things to grow. It may or may not be partly stock transference from England, but more likely, because of graphic location (Down Under is seasonally opposite to the northern hemisphere), a roaring export trade in cut flower cymbidiums raged

Page 76: Bright yellow cymbidiums are the current growers' best sellers.
Above: Showbiz Cymbidium.

Chocolate to raspberry to coconut ice, cymbidiums are represented in almost every colour spectrum.

earlier this century. The trade continues at a lesser rate today.

Boxes and boxes of flowering 'cymbids' (now there's a word never to use) winged their way abroad. Australians also encouraged their plants to grow straighter, shorter stems to better fit the boxes. I'd rather not discuss that one. My tendency is to buy the more billowing, free-form, older varieties.

Above: A lemon-white cymbidium and Chinese calligraphy.

They're cheaper, but without as many new colour choices.

Take care as you read 'how-to' books on orchid cultivation. As cool growers, cymbidiums still need warmth in European winters, just as they have difficulty flourishing in humid equatorial conditions. Always, it is better to make friends with collectors in your area to determine what suits you best. Wasted hot houses are sad sights and environmentally unsound. Heating bills can become unbearable and you may end up cursing the object of your affection.

Appearing exceedingly aloof, these angelic open faced orchids grace five-star foyers around the globe – yet they flourish equally well in modest domestic settings. Grace Kelly wrapped in Monroe's sex appeal. The princess and the showgirl. When it comes to tonal values, think Spice Girls meet La Stupenda.

Chocolate to raspberry to coconut ice, cymbidiums are represented in most every colour spectrum – except bright clean reds. Fresh daffodil yellows contrast with pale honeysuckle apricots. Loquat and persimmon. Terracotta and London tan. Delicate gardenia, fragile magnolia. A host of tender greens includes emerald and jade. In parts of China, the orchid is known as the Jade Lady. Jade having as many color nuances and the same waxy feel as the blossom. Jade also represents good fortune and that equates with the stature of the cymbidium.

The wildest cymbidium I ever encountered was growing in a shade house under the

Above: Cymbidiums at the bi-annual Australian Cymbidium
Society National Show, 1998.

Top: A face is a flower.
Above: Cymbidium detail.

verandah of a house on stilts in Jurien – north of Perth in Western Australia – on the way to Broome. It was a chocolate, pink and red peppery number and positively screamed to be photographed in the otherwise quiet greenery of the shade house. The long range lens did not give enough detail, so I crept closer – stalking the orchid. Through a little wire gate, past some fruiting pawpaws, down the barrel of the macro lens and straight into the heart of the flower.

A piercing 'Bonjour, mate!' shattered the calm morning air and completely shot my concentration. In sheer panic I was ready to run. Orchid speak? No. A talking black cockatoo acting as watchdog!

The lady of the house came out and smiled. 'We trained him to talk in French to scare intruders.' It worked. No more pictures without permission and – heaven forbid – no more talking parrots.

Once in bud the sprays of opulent blossom continue for months on end. Value for money. But here's the catch! What do you do when the flowers drop off? Dump the plant? Move it behind the back shed?

Imprisoned indoors the de-flowered maiden is usually housed for about six months in vain hopes that buds will miraculously re-occur. Not so. The dusty plant fades in the half light and grows sickly. We forget that plants, especially orchids and bonsai, are outdoor creatures. Neglect. Over-watering. The plant goes down for the final count right about the time its owner

Opposite: Deep colours intensify if budded in strong sunlight. Light colours succeed best if bloomed in shade.

decides to give it an affectionate nickname –
then dump it on some unsuspecting gardener.

**Even at the risk of losing a friendship,
do not accept a gift of sick cymbidiums.**
Start with healthy plants. Grow cymbidiums
in light. Under a shady gum tree is ideal.
Thick foliage trees darken the day. The plant
will grow, but never flower. Plant the orchid
in a bucketful of orchid compost (go see your
nurseryman) and with the normal amount
of care and attention any garden deserves,
the little stalwart could well do its stuff for
years to come. Some gardeners do as well
in window boxes and on balconies, providing
wild winds don't prevail.

Colours tend to drift in open areas. There
is a rule of thumb. Reds and deep colours
darken and intensify if they are grown in
strong sunlight until buds open, then they
need shade to keep the colour. Greens
actually turn yellow if exposed to full
sunlight. Light colours succeed best bloomed
in shade and taken to bright light to soften
the colours once the buds are formed.

Right: 'Bonjour, mate!' A French-speaking black
cockatoo watchdog in Western Australia.

I am a camera

A camera is a paint brush, a pen, a piece of charcoal,
a scribble pad to jot notes on. It's the medium. Just remember,
you're the one sending the message.

You don't have to be a whiz at theorems and mathematical equations to take good pictures these days. Trust the computer in your camera, don't try to beat it. Learn to use computers to your advantage.

I don't much believe in 'talented' people. There's only practice, practice and still more practice. Persevere and the breakthrough comes sooner or later. Athlete. Buddhist. Artist. Journalist. Orchid enthusiast. All it takes is drive, energy – and maybe some of Edward De Bono's *Lateral Thinking*. That part about not going up and down in the one hole. Lateral jumps often reveal the solution by giving a different perspective. Your brain receives added information with which you make sideways jumps, to get a different view of the situation. It's the same with setting up shots. Taking good pictures is not the problem. Laziness is.

I'll do anything to get the picture. Bully, clown, coax, entice, support, promise – except intrude.

Kneel down, lie down, stand on your head. Don't expect your subject to do the hard work. Flatter, cajole, excite, entice. Do anything to make a face or flower relax and relate. Forget everything except the energy flow between you and your subject. Here's what someone thought I was up to in Burma,

and I quote verbatim: 'I have enclosed three slides which I'm sure you don't have since your marvellous hulk is in them. We're especially fond (in fact almost infatuated) of the one of you, on your knees, capturing the glorious colour of some Buddhist robe. One of our happiest memories is of the somewhat eccentric companion in our threesome who, at any moment, could be seen lying full length in the Burma sod to get just the right photo; or ignoring an ancient palace to immortalise a weed growing from a crack; or even standing atop a bus for a close-up of a flowering tree …'.

I'll do anything to get the picture. Bully, clown, coax, entice, support, promise – except intrude.

Page 86: Quan Yin, Chinese goddess and Buddhist protector of women.
Opposite: The author.
Above: Life-sized carved marble Buddhas in Mandalay, Burma.

Get a macro lens and a camera that works on automatic. Buy a fairly expensive body.

Avoid arty farty attitudes.

Get yourself a good alarm clock and use dawn light to photograph flowers to advantage. You may have to wait a bit for flowers to pop. Put the camera away after 8am in Australia, or be prepared to waste a lot of film. Those bright blue Aussie skies fool a lot of people. In England, you can shoot much further into daylight, so set your alarm accordingly.

Black and white sit at opposite ends of the spectrum. The prettiest colours are in the middle tones. Go out in the midday sun and you'll record heavy shadows (black) and bleached out (white) colour. All the soft colours are squeezed up in between – except there is no in between. Just glare and shadow. **Go in soft light** (dawn, dusk, slightly clouded conditions) and you will find all the colour you want. So will the camera.

When the sun is behind your subject, expect black silhouettes of your intended, and a blinding, bleached-out background. This is a form of back-lighting – often not the effect you want.

Don't listen to others on location. Hand them a camera to go take their own photos. Concentrate. Take the picture you want. Express yourself.

Show your results to photographers whose work you respect. Listen to what they say, and learn. Write down the bits of information you're going to keep. Chuck out the rest.

Don't flash. Turn off the camera flash whenever possible. The sure sign of an unskilled amateur is over-exposed happy snaps. Don't bore people with bad pictures. Weed out your flops and your portfolio will present a more professional you.

Keep hands free. Wear a shirt or jacket with plenty of pockets to carry spare film, bits of tie wire, a small misting spray in case (heaven forbid) you want dewdrops on your subject. A small square of silver foil, folded to fit a pocket or camera bag, is also useful to reflect a soft highlight back into your subject.

A light misting does make the world of difference to your composition. Water unites the dry woods and earth around your subject. It brings a oneness to the photograph and gives the flower radiance. It makes leaves shine and can add that elusive mystic touch people find entrancing. Hollywood cameramen used Vaseline on the camera lens to soften wrinkles on ageing film stars, but let's not go too far. Shooting through chiffon is another trick, but I've never tried it.

Make sure eyes register or the central part of the flower is the focal point.

Include kids and old folk in your compositions. Smooth skin and interesting wrinkles make fascinating photos. Gnarled logs make good contrast against fresh-faced flowers.

Depth of field. As long as one bit of your photograph is in focus (hopefully the thing you're featuring), it qualifies as an OK picture. To my mind, the click clique are far too obsessed with 'Depth of Field' where every damned thing in the picture has to be in sharp register. Every blade of grass. Every distant hill. Every stone unturned. Harsh focus has nothing to do with the ways of nature.

Try a simple eye test by looking at an object close to you – the background softens. Now look into the distance and see how things up close tend to blur. Your photograph should have that same easy feel. Don't try to get everything in perfect focus if you want your flowers to seem natural.

See yourself as an artist and follow the feel of the Impressionists. Monet's waterlilies are a good starting place. Subtle highs, lows, accents and diffusion. Where's the hard focus. Where's the political correctness. Drop it.

Learn what every camera button does, but select only the ones that work for you.

Don't dwell on compliments. You'll get a swollen head. At the same time, don't spend any time around negative people either (you know if you like the picture in the first place). Keep your opinions of your work to yourself. People may use them against you when you become famous.

Framing in camera. Work those lenses and camera angles until your picture sings. Don't have odd bits popping into your field of vision. Make sure your picture is nicely framed in the lens. Sure, you can cut bits off after the event, but the effect is always amateurish. Not a lot of photographers can frame properly in camera. Waste some rolls learning to get it right. You'll save money in the long run. And try a few rolls of transparency film rather than happy snap prints. Project your transparencies to music. You'll love the effect. Pictures always look better when shown to music.

Start small. The wide angle lens comes last. Don't attempt to capture big scenes at the beginning.

Cameras (and computers) are overloaded with things you don't need. Learn what every camera button does, but select only the ones that work for you. The guy who uses them all is a technocrat – and probably boring. You're into composition and creativity. Skip gadgets and gizmos. Let the picture sing.

Shy violets are hard to find in photos, yet viewers do want to see the wood from the trees. Whether it's faces or flowers, it's good to reveal something. Smooth skin, the curve of the neck, a bit of leg. Show some 'eye candy' – it never harmed any picture. Likewise the flower photo. In tight, up close and let's see the gorgeous contours.

Following pages: Hybrid white cattleyas, Queen of the Orchids.

cattle-lay-ya
don't even think it

Even florists mispronounce *Cattleya*.

Get it wrong and you've missed the boat if you're

about impressing those in the orchid nod.

The most mispronounced flower name in the world is decidedly the most popular orchid in the world. *Cattleya* is named after one William Cattley, a keen orchid fancier in merry old nineteenth century England. Pronounce his name – Cattley – then just add an upsounding 'uh'. Even florists often mispronounce *Cattleya*. Get it wrong and you've missed the boat if you're about impressing those in the orchid nod.

So many species of the plants are named after their discoverer. And that's a good way to start getting a handle on enunciating some of those preposterously difficult plant names. It pays dividends to discover the derivation of your favourite orchid right from the start.

As colourful as cattleyas are, they are often crossed with hardy laelias to introduce even more vibrancy into the equation.

I'm the worst offender, having never studied Greek. The thing that irritates me is the shortening of names (like phalaenopsis down to 'phally') but it's quite 'the go' in orchid circles. It seems to disgrace the flower. Cymbidiums shorten to 'cymbids'. Cattleyas to 'catts'. Maybe catt is acceptable. Catts are rather household pets. 'Putting the catt out' does sound a bit off. Cat doors? Nope. I'll stick to cattleya and learn to say it correctly.

To many people, it's a fight out as to whether orchid means *Cattleya* or *Cymbidium*. To me, both have been ravished by over-enthusiastic hybridising as growers grab for show-winning specimens. We're beginning to understand that the species is most desirable. Breeders are much more inclined to return to nature's beautifully coloured and well-formed originals. Beginners usually start with a few catts (now it's getting to me) or a *Cymbidium* or two. But I will never overcome the horror of those stitched-up corsages from the 1950s.

My preference is for the catts' cousin, little *Laelia*. As colourful as cattleyas are, they are often crossed with hardy laelias to introduce even more vibrancy into the equation. Yellows, orange reds, glorious coppery bronzes. *Laelia tenebrosa* is a favourite. Copper and rust, purple and magenta. What a combination! The very essence of Brazil, its home territory. In Australia, down under, it blooms just before Christmas. Experts tell me that's late. Then everything in my garden is a little behind schedule.

There's also a stunning scarlet-and-cerise hybrid called *Slc Jewel Box*. The *Slc* is short for *Sophro-laelia-cattleya*. And Sophro is short for *Sophronitis*, a small Brazilian orchid which can be used to contribute a rich redness to a *Cattleya* hybrid. Hence the magenta of the *Cattleya* is underscored by the brilliant red of the *Sophronitis*. But there is no need to get too involved. Just ask for a *Jewel Box*. I grew one outdoors in the bush house for years.

Page 94: One of the lovely white cattleya hybrids.
Opposite: Brilliant Brazillian *Laelia tenebrosa*.
Following pages: *Slc Jewel Box*, a 'Sophro cross'.

And another intensely red-orange, small flowered chap that came from the tropics, without a name. I'll take a guess at *Laelia cinnabarina*.

Brassavola is another genus in the *Cattleya* tribe, and it's this group that contributed the showier full lips to *Brassio-cattleya* orchids. Then there is *Epidendrum*, from which the enormously popular crucifix orchids spring. Most gardens feature a clump of Crucifix, and if not in yours, may I suggest a start.

The whole subject of the *Cattleya* tribe is best studied in a highly respected book written in the 1950s by Rebecca Tyson Northen, *Home Orchid Growing*. The famous American author begins with sound advice: 'Cattleyas have many relatives that are charming and delightful in themselves. Among them are *Epidendrum*, *Laelia*, *Sophronitis*, and *Broughtonia* ... the handsome cattleyas, the most showy members of the tribe, should not be allowed to overshadow their lovely cousins.'

I took the hint!

As cut flowers, cattleyas are featured up front in a florist's window. Other blossoms may be used for bulk and background, but you'll find showy cattleyas (and the folk who grow them), up close – and personal.

Right: Cattleya *amethystoglossa*.

the lost slipper

A Cinderella story

Near the end of the nineteenth century, the Botanic Garden of Penang shipped a ravishing little slipper orchid to England under instruction of Charles Curtis, a collector for the famous firm of Veitch of Chelsea.

The account of Curtis' orchid did not, at first, appear to have a happy ending. His little Asiatic import eventually expired from excessive care and attention, and a search began to find another of the same species, *Cypripedium curtisii*. For over 50 years the search continued. No success. Then (and here's the Prince Charming part), a young Swede named Ericson found a portrait of the elusive slipper in a mountain retreat when he was orchid hunting in Sumatra.

The painting was inscribed *Charles Curtis' contribution to the adornment of the house.* The collector suddenly realised that everyone was pursuing the little Lost Orchid in all the wrong places. Orchid hunters repeatedly left false trails to fool other collectors, in order to protect the locations of their recent finds.

The Swede hunted high and low for weeks and was about to abandon his quest when, underfoot, he spotted the long lost slipper sheltered deep in shady rock crevices. The lost orchid was found. But that's not the end of the story.

Orchids are forever being renamed and reclassified as experts discover more about their kinship. Today, should you look for *Cypripedium curtisii*, you'll also find it hard to uncover. The name has been changed. Now you look for *Paphiopedilum curtisii*.

For me there is still a mystery to unravel. What became of the lost art work last seen hanging on the hut wall in Sumatra – the key to the rediscovery of Curtis' lost slipper?

Opposite: I still prefer the simple slippers
to the saucer-shaped show-winners.

Following pages: Tiger! Tiger! Burning bright
in the forest of the night …

the tiger and the
dancing
ladies

On my first visit to Raffles in Singapore, I was bowled over

by an immense centrepiece of yellow and gold orchids

gracing the doorway to the hotel. Speckled in sunlight . . .

masses of tiger orchids and dancing ladies.

The long branching sprays of seductive little yellow flowers waltzed over a thicket of vibrant striped tigers, all about ready to pounce. Here and there, an occasional spider orchid popped out, presumably to keep the ladies on their toes.

Having enjoyed a couple of gins and quinine water (and sampled one or two Singapore Slings, the house specialty), I decided to front again the following day with camera, to shoot the orchid arrangement. No luck! The stunning feature changed daily and had been replaced with a snow storm of coconut ice vandas, *Miss Joaquim*, the flower symbol of Singapore.

Bang! Dead tiger. They didn't muck around in those days. I wasn't game to mention the dancing ladies, because the old prostitute houses were only a stone's throw away from the hotel.

'Where'd the tigers go?' I chatted up the nearest porter.

'Inside hotel.' He led me deeper into the dark. We arrived at the original pool room. He pointed to the corner. 'Tiger shot under

Page 106: Dancing ladies on a Kentia palm, author's garden, Darlinghurst.

Above: Dancing ladies, Oriental Hotel, Bangkok.

billiard table.' Legend has it a striped tiger wandered into the bar one night and gave the patrons a bad time. Bang! Dead tiger. They didn't muck around in those days. I wasn't game to mention the dancing ladies, because the old prostitute houses were only a stone's throw away from the hotel, and if they once shot tigers in the hotel, who knows what could happen to real 'dancing' ladies!

Ask to see a tiger today, they'll show you an *Ondontoglossum grande* (now recognised as *Rossioglossum* in the latest orchid manuals). Men are forever re-classifying orchids as more and more is revealed about the species. Ask to see the old prostitute houses, and locals appear not to understand. 'No, no. No prostitute house in Singapore.' The upmarket metropolis has cleaned up more than her streets. Enquire after dancing ladies and you'll probably get a ticket to a ballet recital.

Swinging north to Bangkok, I discovered dancing ladies by the drove in the Writer's Wing of the salubrious Oriental Hotel, on the River of Kings. The private apartments were used to billet authors who wanted to get away from it all, and the visitors' books in the library read like a literary Who's Who. Joseph Conrad, W. Somerset Maugham, James A. Michener, Yukio Mishima, Noël Coward. A white staircase rose in the middle of the vestibule where guests took tea with cucumber sandwiches or gently sipped the occasional Pimms. The dancing ladies? They shone in great branching arrangements from the mezzanine above, allowing the celebrated guests to peer through and observe the talent incognito. Dancing ladies are really useful orchids.

Thai dancing ladies are also about as local as Gordon's London Special Dry Gin. Born to dance, these babies are also bred to travel. They're sympodial, meaning a new stem develops each year; the next lead branch takes over when an old one flowers each season. Each stem has a swollen, bulb-like region called a pseudobulb. Sympodial also means the plants walk around a lot because of their growing habit. A healthy specimen will outrun its container every year or so. I tie them to trees so they can go where they choose.

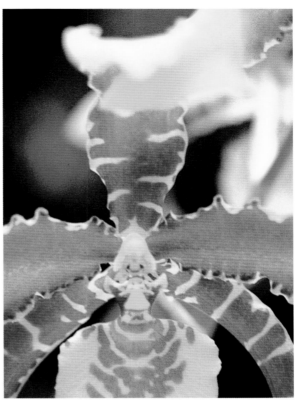

Above: A tiger is a *Rossioglossum*.
Following pages: *Phalaenopsis* – moth or moon orchid.

Dancing ladies are as South American as Carmen Miranda, as are Singapore's flowering tigers. Westerners enjoy oriental myths and legends – and Asians, being great storytellers, have developed the tiger and dancing lady routine to a fine art. If you want to see *Oncidium, Ondontoglossum, Rossioglossum, Brassia* (spider orchids) and a legion of their tropical relatives in their native setting, head south of the border, down Mexico way – to Guatemala, Florida, the West Indies and Central America.

moon orchids

My favourite species is called *Phalaenopsis* and is native

to the Philippines. Aphrodite, orchids in the moonlight.

No wonder moon orchids grace every

grand piano from New York to Timbuktu.

Enchanting moon orchids can make expensive playmates. Big city shops make a killing. For instance, flowering plants in New York almost cost me an arm and a leg.

Like Peter Allen, I was caught 'somewhere between the moon and New York City' in a holding pattern when my appendix burst. Two weeks later and 14 000 American dollars poorer after the operation, I was discharged, and lobbed back into the thick of things at eight o'clock on a Monday night. 'Yep! You can go.' Some enchanted evening.

The following day I found myself temporarily residing in a small East 73rd Street apartment, waiting for stitches to be removed – and still determined to find a New York publisher for an Asian book idea I was peddling. I'd come so far. The book on the Philippines was positively swamped with shots of indigenous moon orchids. The Burma dummy burst with Buddhas and soft cane *Dendrobiums*. Malaysia, *Vandas*. But *Phalaenopsis*, lovely Filipino moon orchids, were to be my calling card.

The doorman at the apartment block sent me crosstown to a market where *Phalaenopsis* supposedly sold at a fair price. Shocked, I returned with a couple of less-than-salubrious specimens in pots – 240 American dollars apiece. Some enchanted orchids!

New Yorkers set up stings to catch cons at their own game, but Australians are better at spinning yarns and telling tall stories. My telephone pitch was intended to bring hard-nosed publishers to the apartment rather than lug the heavy

Page 112: The orchid lip has a beauty of its own.

Opposite: A pink haze hybrid.

Following pages: Hybrids also do creative things with spots and stripes.

book presentation around town. And it almost worked.

The orchids, rented audio visual equipment, a pitcher of martinis and platters of sushi from the downstairs Japanese restaurant were the call birds. Friends rallied as snow fell and freezing cold set in. A furtive glance at the yuletide decor and Macy's annual Christmas parade and I was out of there – wearing a banker's grey cashmere coat bought on sale at Bloomingdale's. Back home to Sydney to a much warmer climate. I never needed to wear that damned overcoat again.

Today, a single plant stands just inside the kitchen window beside the hot water system. It loves steam and heat …

My little inner city hot house was waiting. As year-round warmth is needed to grow *Phalaenopsis*, getting the orchid climate right was my magnificent obsession in those frustrating embryonic years. Today, a single plant stands just inside the kitchen window beside the hot water system. It loves steam and heat, and it is fun to watch as the flowering procedure progresses. But that feisty little hot box was crucial to my orchid learning curve way back when.

The juxtaposition was hilarious. Inside the garden, hot house orchids and a sub-tropical jungle. Out in the street, an old sandstone prison and rows of dwellings housing pimps, prostitutes and alternative lifestyle *glitterati*.

Also called 'moth orchids' (Greek, *phaluna* meaning moth), appreciation of *Phalaenopsis* goes deeper than people sometimes care to admit. It's the *luna* part of *phaluna* that gets me going. Luna, the moon. My garden is planted according to moon cycles. And moon is the look of the famous white *Phalaenopsis*. My favourite, a species from the Philippines, is called *P. aphrodite*. Aphrodite, orchids in moonlight. Who cares about hairy moths.

At first sighting in a shop window, my nose is pressed to glass. Even scatty specimens beckon to be adopted and rehabilitated. The price of a plant may seem high at first. Double the cost of this book, and then some. But what returns! Hours of pleasure, sheer economy. My current purchase flourished for four months. Gorgeous white blooms on two long branching stems, about fifty individual flowers. It worked out at 65 dollars for 120 days, about fifty cents a day. By carefully pruning the stems after flowering, and prudent use of weak fertiliser in warm weather (starve the plant in the kitchen through winter), another season. Two hundred and forty days, 25 cents a day. This year it could happen again.

Look, just go buy someone a *Phalaenopsis*. Let hearts flutter. Moon orchids do a power of good. Ask Aphrodite, the Greek goddess of love.

Opposite: *Phalaenopsis lueddemanniana*. Glossy little butterfly.

Following pages: On the Big Island of Hawaii, an orchid house the size of a small playing field.

the noble
hawaiians

The story unfolded that Japanese companies were cloning big time in Hawaii, churning out *Dendrobium nobile* (no-bill-ee) sixteen-to-the-dozen.

For years I imagined Yamamoto was an orchid species endemic to Hawaii – a rare *Dendrobium* perhaps. My ears pricked up every time someone mentioned 'the Yamamotos from Hawaii'. The words intrigued me as did reference to 'clonal culture'. A famous family name? Now we're getting close.

The story unfolded that Japanese companies were cloning big time in Hawaii, churning out complex hybrids from *Dendrobium nobile* (no-bill-ee) sixteen-to-the-dozen. One would turn up every so often and clean up at orchid shows. The Japanese breeders even learned to extend the plant's bloom cycle, intending to put paid to the popularity that *Cymbidium* and *Cattleya* had previously enjoyed. It transpired there was room enough for everyone.

The primary *nobile* I know is pinky-mauve, with white to lemon-cream interior and a Black-eyed Susan effect punctuating the centre spot of the flower. Quite outspoken! The white (var. alba) version is more gentle in manner, but both are needed to bulk out any backyard orchid collection. They act as call birds bringing your eye down to the more diminutive dendrobiums. Each flower is the size of a small eggcup, and a dozen or more sprays of blossom can shower the plant at once, in springtime.

Native to far eastern countries such as Burma, India, Thailand and Indo-China, soft cane Dendrobiums grow on forest trees from lowlands to quite cool highlands of 4000 feet elevation. They are extremely hardy; surviving warm, hot, and downright freezing temperatures. In Hilo they found Nirvana. Heaven in Hawaii.

Growing *nobile*-type dendrobiums cloned in Hawaii began to appeal to me. But why buy locally from limited stock? Get to the source. Photograph the flowering phenomena and bring home the plants as a bonus. After discovering how complicated importing plants can become, the plan seemed a bit knotty. However, tangled web or not, the decision was 'all systems go'.

The arrival in Honolulu was chaotic. The economy flight had set down in every pit stop across the Pacific. In Honolulu, nothing much remained to remind me of previous visits. Big Mac had finally overpowered the smell of white ginger blossom and heady frangipani leis. For old times' sake, a couple of Mai Tais at the Royal Hawaiian and I was out of there, headed for the Big Island of Hawaii where my orchids waited. Two laid-back nights in Hilo, then an ill-fated decision to hire a car and drive solo to the famous nobiles. A few wrong turns and the next night was spent up in the National Park – at the time suffering enormous volcanic activity.

Standing with the crowds, I watched all hell break loose. Roads blocked. Rivers of red lava. News journalists screaming around, and everything pitch black until the next eruption. Then it developed the orchid farms were back at Hilo where I had come from.

'They're on the other side of the island.'

'Thanks, mate.'

Needing a navigator for the 120-kilometre night drive back to Hilo, I picked up a giant

Page 122: The loot brought in from Hawaii flowered in the first year.

Opposite: *Dendrobium nobile*-type hybrid.

Hawaiian. A strong silent islander whose stature suddenly appeared menacing as he entered the car and overflowed the front seat. The heavy smell of a day's work in the canefields prevailed, and he carried the largest machete I'd ever seen. It was like an old *Road to* movie except there was no Dorothy Lamour and this was no laughing matter. Winding road, rocky coast, pounding surf, dark night. In the luminous green light of the dashboard, the scene played like a B-grade horror film crossed with 'The X Files'.

Street lamps never looked as good as those on the outskirts of Hilo around 3am. Fitful sleep, then off on the road to the orchids. This time in the right direction.

Driving through macadamia plantations at sun-up set the mood of the next day. Open countryside, a few stop-overs at roadside orchid establishments, then a sudden panorama – a hillside rolling in orchids, arranged shed by shed, plant by plant. A security gate, chat with the Japanese manager and his pretty wife, and my camera went berserk. An orchid house the size of a small football field, a riot of *nobile*. No need to choose subjects. It was as easy as shooting fish in a barrel. The light in the open-sided glass-topped pavilion was perfect. No false fluorescent lighting effects, just soft free-falling daylight flooding in from every angle. A photographer's dream!

Where young plants were not growing in totally open conditions, large upright fans were used to create breezeways. The breeze from the fans kept things at just the right temperature both night and day. In Hawaii, the adult plants need no shade at all. They grow all year round in the open. There is a direct relationship between temperature, light, and watering.

My garden in Sydney is another matter. Cold July and August wind can be quite fierce on plants out-of-doors. Still my little collection of soft cane dendrobiums struggles through and comes up smiling in September and October. *Phalaenopsis* are a different matter (that's where glasshouses come in). My moon orchids move indoors during winter. Pretty much a replay of medieval times where family slept with their prize livestock. One plant in the kitchen window beside the garden; a couple hanging in the annexe upstairs beside the loo where the sun shines all day. A few scattered around the bedroom, near daylight. Orchids everywhere.

The shock of leaving paradise, being quarantined and then ending up in a garden with less pedigreed stock may have been too much for the nubile little *nobiles*. The neighbour's tree (a brute of a thing from east coast USA) stole their light as one by one they carked it. The original Black-eyed Susan variety and its lovely white alternative show greater tolerance, and their numbers multiply as they reinvent themselves each year. More pups. More keikis. More plants.

The tender, well-cared-for Hawaiians just weren't plucky enough, though I get to visit one of the imports in a friend's garden where it hangs on a side fence in full searing sun. A brilliant yellow soft-cane called *Dendrobium*

Above: *Dendrobium Shinonome* 'Compact', soft cane
nobile bred by Yamamoto, Hawaii.

nobile Pittero Gold 'Grace'. Now that new owners have removed that damned tree near my garden, I may get a pup from my friend's plant and reintroduce Hawaiian soft-canes back into my garden. Or maybe face another trip to Hawaii to start all over again.

A deciduous species, *nobile* does not lose its leaves until the second year. The naked canes are what produce the bracts of flowers in the second winter as you rest the plant in a dry spot. Do not continue feeding in cool weather, or you'll encourage loads of little new plants on the side of the stems where you were expecting blooms. These babies are called keikis (kee-kee). Some refer to them as pups. The single keiki can be eased away from the mother plant when it has developed its own root system. Planted in loose bark, and fertilised regularly, the keiki grows to a fine upstanding specimen in a few seasons.

As universal debate centres around whether cattleyas are the most widely grown of all orchids in the world, many books give instructions like 'handle as for cattleyas' or 'grows in Cattleya conditions'. Take the hint and you'll find the adaptable little *nobile* that journeyed from Indo-China to clone in Hawaii can hold its own in almost any company.

Opposite: Looks like Black-eyed Susan. It blooms outdoors with a minimum of attention.
Following pages: It is sometimes called *D. virginalis*.

pathway to the flowers

A terrace of orchids here. A hanging garden there.
An extravagantly flowered foyer in some swank Hong Kong
hotel. Orchids blooming with gay abandon with
no apparent idea of the joy they are spreading.

Some of the world's most beautiful sights are waiting for you in this part of the planet. Draw an arc, a kind of pivot across the Pacific – Sydney to San Francisco. Now sweep it across to Asia, Japan and down through China to Burma, then back down under Australia to rejoin Sydney. You've circled my home territory where I travel most, with camera and sketch pad. While these accounts are just one man's view of orchids, the photographs and storylines may serve to give you a window into what propels us along the pathway to the flowers.

Everyone daydreams. My love affair with flowers and people in exotic places began at an early age. Robinson Crusoe was an early hero of mine. A special edition of *The Life and Strange Surprising Adventures of Robinson Crusoe* was given to me by my parents for my eighth birthday. Gold embossing dominated the sedate burgundy cloth binding, and pressed between those pages were some pretty rollicking events. Storm. Shipwreck. Mutiny. Cannibalism. None of that Biggles tomfoolery for me. Deserted islands were to be my bag with Crusoe, his parrot, Man Friday and do-it-yourself grass huts – with black and white etchings to visualise in colour. And horrible pirate tales to set my mind racing. Then cinema took hold as Bloody Mary and *South Pacific* caught the imagination. Finally, when Cyd Charisse came out of the mists of *Brigadoon* I was headed for fantasy land. Goodbye Mary. Goodbye Crusoe. The search for the unattainable flower was on.

Twenty-something years later the dream began to materialise: Hawaii, Penang, Hong Kong, Manila. Finding destinations was not a problem. Pussyfooting around departure halls was the big turn-off. Airports and pushy people horrify me. Long drawn-out inflight periods. Dreadful food – which I ate. Baking heat on the tarmac in Calcutta. Cold drizzle and cancelled midnight flights in Beijing. Attempts to recall the exact lyrics of a Noël Coward song to pass the time: '. . . why do the wrong people travel, travel, travel, when the right people stay at home.' People looking at the strange foreigner who sang badly.

Avoiding tourists with cameras is now my *modus operandi* … Solo is my trip. Me, my camera, and the flower I'm about to meet.

Then an arrival, a new experience, a funny incident. The travel bug would bite again.

I was late getting into photography, choosing rather to send back sketches and letter/diaries when on the move. Avoiding tourists with cameras (and trying to hide mine) is now my *modus operandi*. There's nothing as boring as a chatty camera buff en route. Solo is my trip. Me, my camera, and the flower I'm about to meet.

Page 132: An orchid pathway, Oriental Hotel, Bangkok.
Opposite: Singapore orchids, original Raffles hotel. All gone.
Following pages: Vanda Miss Joaquim.

Some marvellous images of orchids crossed my path in unexpected places. A terrace of orchids here. A hanging garden there. An extravagantly flowered foyer in some swank Hong Kong hotel. A *vanda* nursery in the foothills of distant mountains. A mixed grill of poor man's orchids struggling for life in the back window of a run-down house. A state forest. Elbowing through crowds in posh shopping centres. Dendrobiums growing through thatched roofs on remote village huts. Precious wild orchids underfoot; step lightly. And close up, as always, orchids blooming in gay abandonment with no apparent idea of the joy they are spreading.

In Burma, I travelled the pathway of flowers to find my beloved dendrobiums and blue vandas. Later I realised I was also following the way of Buddha. Burma into Thailand. Through China. On to Japan. No matter what political persuasion a country adopted, Buddhism endured. The lotus is the flower at the foot of Buddha, but love of orchids lies deep in the heart of every Oriental. For me the two flowers are inextricably linked.

> Botanic gardens are my translation of 'tell me what you have eaten and I'll tell you what you are'. One glance and you discover much about the community spirit.

To understand the people, my first undertaking is to visit local botanic gardens on the day of arrival. One glance and you discover much about the country. Climate. Bird life. Community spirit. How the city wakes up. Gardens are my translation of the much quoted 'tell me what you have eaten and I will tell you who you are'. Bad vibes and I would move on. But I haven't met a botanic garden yet that didn't fascinate.

In her book *Orchids of Papua New Guinea*, acclaimed botanist Andreé Miller locates the orchid's habitat quite distinctly: 'The greatest concentration of this orchid, *Dendrobium chrysoglossum*, is in the Casuarina trees on the banks of the Ramu river.'

Above: *Dendrobium chrysoglossum.*

Opposite: The healthiest orchid I've ever seen growing.
A virile blue Vanda, near Bangkok airport, Thailand.

There are reckoned to be about 25 000
to 30 000 orchid species. They grow from
3500 metres up to sea level, from the equator
to the Arctic Circle. You must search out
orchids to love them. They do not freely
come to you. Joseph Conrad said it better:
'Going up that river was like travelling back
to the earliest beginnings of the world, when
vegetation rioted on the earth and the big
trees were kings ...'

Orchids are not as easy to love as, say,
roses or daisies. OK, so someone gave you
an orchid as a gift. The message is that you
are very special to them and they are saying
it with orchids. But there is more to it than
that, if you get my message. The relationship
deepens as you get to understand the nature
of orchids. You need to work with them to
really love them – and that can be frustrating.
I photograph more and collect orchids
less. But my garden is still full of the easy
growers. Now I go searching for the rare
ones I don't possess.

This book is intended to give you a leg up
the orchid ladder. In layman's terms 'it's the
journey, not the arrival'; 'getting there is half
the fun' and 'one first step . . .' The good
thing about clichés is that they work.

Just go find your orchid.

Right: Orchids in air, thriving on a decaying tree stump
in Asia's Golden Triangle, Northern Thailand.
Following pages: Colourful Vandas in the
Singapore botanic gardens.

Aerides, 52
Astonishing Stanhopeas, The, 42
Bletilla, 21, 22
Brassavola, 100
Brassia, 109
Broughtonia, 100
Cattlea Little Angel, 61
Cattley, William, 96
Cattleya, 96, 100, 124
    amethystoglossa, 101
cattleyas, 94–101, 128
    white, 92-93, 94
Coelogyne pandurata, 44–45
Cooktown orchid, 23, 60, 72
Curtis, Charles, 103
Cymbidium, 78, 96, 124
cymbidiums, 52, 76–84, 96
    jade, 74–75
    lemon-white, 80
    Showbiz, 79
Cypripedium curtisii, 103
dancing ladies, 107–109
Dendrobium, 51, 52, 53, 114
    antennatum, 21
    bigibbum, 23, 72
    chrysoglossum, 139
    chrysotoxum, 53, 69, 72
    densiflorum, 68, 72
    Ellen, 71
    farmeri, 66, 72
    gracilicaule, 51
    gracillimum, 51
    kingianum, 50, 53, 70
    moniliforme, 18
    nobile, 61, 70, 124, 125, 126,
        128, 129
    nobile Pittero Gold `Grace', 128
    pierardii, 64–65, 70, 72

Shinonome `Compact', 127
speciosum, 8, 10, 50, 70
virginalis, 130–131
dendrobiums, 62, 67, 138
    Indian, 38, 68, 72
    New Guinea, 73
Dolly x Claude, 78
Epidendrum, 100
Greer, Barney, 42
Griffith, Dr William, 29
haiku, 58
Hawaii, 120–124
Home Orchid Growing, 100
Hooker, Sir Joseph, 29
Ikebana, 57, 58
Jade cymbidium, 74–75
Jade Lady, 80
Jake's Folly, 78
Jewel Box, 96, 98–99
keikis, 128
King Orchid, 8, 50
Laelia, 60, 96, 100
    cinnabarina, 59, 100
    tenebrosa, 96, 97
Linne, Carl (Caroulus Linnaeus),
    18, 52–53
Miller, Andree, 138
Miss Joaquim, 108, 136–137
moon orchids, 29, 30, 60, 110–111,
    112–119, 126
moth orchids, 60, 110–111, 118
Oncidium, 109
Ondontoglossum, 109
    grande, 108
Orchids of Papua New Guinea, 138
orchis, 18
Paphiopedilum curtisii, 103
Phaius tankervillae, 16

Phalaenopsis, 51, 52, 60, 110–111,
    113, 114, 118, 126
    aphrodite, 118
    lueddemanniana, 119
photography, 87-91
Queen of the Orchids, 92-93
Renanthera Brookie Chandler,
    14–15
Rock Lily, 8, 50
Rossioglossum, 108, 109
Sarcochilus hartmannii, 46
Showbiz cymbidium, 79
Showgirl `Glamour Jane', 78
Slc Jewel Box, 96, 98–99
Sleeping Nymph, 78
Sophronitis, 96, 100
spider orchids, 108, 109
Stanhopea, 37, 38, 40
    graveolens, 40, 41, 42
    inodora, 38, 40
    nigroviolacea, 34–35, 38, 40
    reichenbachiana, 42
    tigrina, 39, 40
    wardii, 36, 42, 43
stanhopeas, 38, 40
Stony Range, 10
tiger orchids, 104–109
Tyson, Rebecca, 100
Vanda, 28, 38, 51, 52, 114, 138
    coerulea, 27, 29, 30, 33
    hybrid, 26
vandas, 28, 29, 38, 138, 142–143
    blue, 24–25, 29, 30, 31, 138
    red, 19
Vanilla planifolia, 21
Yamamoto, 124, 127
    Yellow Ribbon `Delight', 13